Universal Holistic Philosophy

Copyright Page

This book is copyrighted for 2021

Title: Universal Holistic Philosophy

Subtitle: And How to Understand the Universe

The Crazy and Out of the Box Series Book 5

By Martin K. Ettington

All Rights Reserved USA 2021

ISBN: 9798493130960

Printed in the United States of America

Universal Holistic Philosophy

Universal Holistic Philosophy

Most of us have been trained by schools and universities as to what scientific and historical truths are supposed to be. However, I maintain that much of what science and history teaches us is in error because the proponents of different points of view have too narrow an understanding of the truth that reality presents.

This narrow point of view also limits our understanding of many issues. The theory of plate tectonics is a good example. First proposed in 1915 it took fifty years to find acceptance as scientific evidence was slowly unearthed to support it.

The same thing is true with the questions being asked today about subjects like the paranormal, longevity, ancient history, and much more

I call this new view of Reality "Universal Holistic Philosophy" (UHP) based on the concept that it takes many different pieces of information from diverse subjects and viewpoints to generate real understandings of certain phenomenon.

The purpose of this book is to review some of the examples I've experienced to show how a Universal Holistic Approach (UHP) to different topics can generate these new understandings.

In essence Universal Holistic Philosophy (UHP) is all about combining disparate sources of information and different perspectives to generate creative and out-of-the-box conclusions and theories which may be a lot different than conventional thinking and may shed light on answers nobody has really thought about.

Universal Holistic Philosophy

This book explores many of these concepts and how they came about. First we examine what are the steps to learning how to think in UHP terms. Then are the cases I've experienced of UHP thinking from researching and writing my books.

Universal Holistic Philosophy

Other books by Martin K. Ettington

Spiritual and Metaphysics Books:
Prophecy: A History and How to Guide
God Like Powers and Abilities
Enlightenment for Newbies
Removing Illusions to Find True Happiness
Using the Scientific Method to Study the Paranormal
A Compendium of Metaphysics and How to Guides (Six books together in one volume)
Love from the Heart
The Enlightenment Experience
Learn Your Soul's Purpose
Pursuing Enlightenment
A Modern Man's Search for Truth
Use Intuition and Prophecy to Improve Your Life
The Handbook of Spiritual and Energy Healing

Longevity & Immortality:
Physical Immortality: A History and How to Guide
The Commentaries of Living Immortals
Records of Extremely Long Lived Persons
Enlightenment and Immortality
Longevity Improvements from Science
The 10 Principles of Personal Longevity
Telomeres & Longevity
The Diets and Lifestyles of the World's Oldest Peoples
The Longevity Six Books Bundle
Long Lived Plants and Animals

Science Fiction:
Out of This Universe
The Immortals of the Interstellar Colony
The Mystic Soldier
Visiting Many Universes
The History of Science Fiction and Fantasy

The God Like Powers Series:
Human Invisibility
Invulnerability and Shielding
Teleportation
Psychokinesis
Our Energy Body, Auras, and Thoughtforms
The God Like Powers Series— Volume 1 Compilation

The Yoga Discovery Series:
Yoga-An Ancient Art Form
Hatha Yoga-Helping you Live Better
Raja Yoga-Through the Ages
The Yoga Discovery Package

Business & Coaching Books:
Creating, Paublishing, & Marketing Practitioner Ebooks
Building a Successful Longevity Coaching Business
Why Become a Coach?
The Professional Coaching Success Trilogy
2020-Make Money Writing and Selling Books
The 2020 Handbook of High Paying Work Without a College Degree
The important of Creativity and How to Improve Yours

Universal Holistic Philosophy

Science, Technology, and Misc.
Future Predictions By and Engineer & Seer
The Unusual Science & Technology Bundle
The Real Atlantis-In the Eye of the Sahara
Removing Limits On Our Consciousness-And Thinking Outside the Box

Survival
Survival of Humanity Throughout the Ages
33 Incredible True Survival Stories
The Importance of Fire in History and Mythology
How to Survive Anything: From the Wilderness to Man Made Disasters
Building and Stocking a Nuclear Shelter for less than $10,000
The Human Survival Five Books Bundle

Legendary Animals and Creatures
Are Cryptozoological Animals Real or Imaginary?
Fire in History and Mythology
All About Dragons
Sea Serpents and Ocean Monsters
The Legendary Animals Five Books Bundle

Ancient History
The Real Atlantis-In the Eye of the Sahara
Ancient & Prehistoric Civilizations
Ancient & Prehistoric Civilizations-Book Two
The History of Antediluvian Giants
The Antediluvian History of Earth
Ancient Underground Cities and Tunnels
Strange Objects Which Should Not Exist
More Out of Place Artifacts
Strange and Ancient Places in the USA
A Theory of Ancient Prehistory And Giant Aliens
The Destruction of Civilization About 10,500 B.C.
A Timeline of Intelligent Life on Earth

Aliens and Space
Aliens and Secret Technology
Aliens Are Already Among Us
Designing and Building Space Colonies
Humanity and the Universe
All About Moon Bases
All About Mars Journeys and Settlement
The Space and Aliens Six Books Bundle
The Space Colonies and Space Structures Coloring Book
All About Asteroids
Spaceships, Past, Present, and Future
Astronauts, Cosmonauts, and Other Important Space Flyers
All About Mars Journeys and Settlement
Mining the Asteroid Belt

Time Travel and Dimensions
Real Time Travel Stories From a Psychic Engineer
The Real Nature of Time: An Analysis of Physics, Prophecy, and Time Travel Experiences
Stories of Parallel Dimensions
We Live in a Malleable Reality-and We Can Change It

Universal Holistic Philosophy

The Longevity Training Series

(A transcription of the online Multimedia Longevity Coaching Training Program)

The Personal Longevity Training Series-Book1-Long Lived Persons
The Personal Longevity Training Series-Book2-Your Soul's Purpose
The Personal Longevity Training Series-Book3-Enable Your Life Urge
The Personal Longevity Training Series-Book4-Your Spiritual Connection
The Personal Longevity Training Series-Book5-Having Love in Your Heart
The Personal Longevity Training Series-Book6-Energy Body Health
The Personal Longevity Training Series-Book7-The Science of Longevity
The Personal Longevity Training Series-Book8-Physical Body Health
The Personal Longevity Training Series-Book9-Avoiding Accidents
The Personal Longevity Training Series-Book10-Implementing These Principles

The Personal Longevity Training Series-Books One Thru Ten

These books are all available in digital and printed formats from my
website and on Amazon, Barnes & Noble, Apple ITunes, and many other sites

My Books Website is: http://mkettingtonbooks.com

Universal Holistic Philosophy

Signup for our Mailing List to get the following:

1) A discount coupon for 25% discount on all books on our site

2) Occasional Notices of new books available
3) Occasional Email on other offerings of ours (Monthly)
Go to this link to sign-up:

http://personal-longevity.com/mkebooks/emailsignup/

And click this link to get the FREE 102 page Ebook titled

"Secrets of Many Things"

If you have any questions about this book or other subjects please contact the Author at:

mke@mkettingtonbooks.com

Universal Holistic Philosophy

Table of Contents

1.0 Introduction 1
2.0 Factors Needed for UHP 3
   2.1 Interests in the Unknown 5
   2.2 Science Fiction and Fantasy Are Important ... 7
   2.3 Getting a Good Education 9
   2.4 The Importance of an Open Mind 13
   2.5 Internet Resources 15
   2.6 Expansion of the Scientific Method 17
   2.7 Data Consistency 21
   2.8 Personal Experiences 23
   2.9 Allowing Time for Ideas to Develop 25
   2.10 Diverse Facts and Their Integration 27
   2.11 Thinking like Einstein 29
3.0 Holistically Derived Understandings 31
   3.1 Investigating the Paranormal 31
   3.2 The Nature of Time 37
      3.2.1 The Future has many probable Paths ... 37
      3.2.2 Time Travel Time Warps 43
      3.2.3 Our Core Spirit Sees Time Differently ... 57
   3.3 Spiritual Enlightenment 61
   3.4 Immortality and Longevity 69
   3.5 Ancient History 73
      3.5.1 Younger Dryas Comet Impacts 75

- 3.5.2 The Vulture Stone 81
- 3.5.3 The Flood of the Bible 85
- 3.5.4 More Ancient History Realizations 87
- 3.6 UFOs and Aliens 89
- 3.7 The Multiverse 91
- 3.8 Human Will and Survival 97
- 3.9 Legendary Animals and the Multiverse 107
- 3.10 Our Malleable Universe 109
- 4.0 Directions for Education 111
- 5.0 Summary 115
- 6.0 Bibliography 117

Universal Holistic Philosophy

1.0 Introduction

Most of us have been trained by schools and universities as to what scientific and historical truths are supposed to be. However, I maintain that much of what science and history teaches us is in error because the proponents of different points of view have too narrow an understanding of the truth that reality presents.

This narrow point of view also limits our understanding of many issues. The theory of plate tectonics is a good example. First proposed in 1915 it took fifty years to find acceptance as scientific evidence was slowly accumulated to support it.

The same thing is true with the questions being asked today about subjects like the paranormal, longevity, ancient history, and much more

I call this new view of Reality "Universal Holistic Philosophy" (UHP) based on the concept that it takes many different pieces of information from diverse subjects and viewpoints to generate real understandings of certain mysterious phenomenon.

The purpose of this book is to review some of the examples I've experienced to show how a Universal Holistic Approach (UHP) to different topics can generate these new understandings.

I would like to think that in the over 100 books I've written on a variety of subjects that I've been able to develop and

Universal Holistic Philosophy

propose some innovative ideas and theories and that I was able to do so by utilizing Universal Holistic Philosophy.

In essence Universal Holistic Philosophy (UHP) is all about combining disparate sources of information and different perspectives to generate creative and out-of-the-box conclusions and theories which may be a lot different than conventional thinking and may shed light on answers nobody has really thought about.

This book explores many of these concepts and how they came about. First we examine what are the steps to learning how to think in UHP terms. Then are the cases I've experienced using UHP thinking from researching and writing my books.

The realizations developed from this work are pretty unique in many cases too.

Universal Holistic Philosophy

2.0 Factors Needed for UHP

One of my biggest concerns today is that young people are grow up living in "Mental Straightjackets" by being taught things which are partial truths or truth being obscured by societal assumptions which may be wrong or partial truths in many cases. There is a real need for more broad minded learning and a better habit of skeptical inquiry for things we are taught in schools, universities, and society in general.

There is also the problem that many young people are taught "politically" correct science which may not be backed up by facts. I site "Manmade Global Warming" as a good case in point. Without discussing the whole issue here there is a distinct lack of science in discussions about this while many scientists and politicians have taken positions based on their funding interests. Also the fraud which exists in altering fundamental temperature data has distorted an objective view of this subject.

My approach to learning and seeing the world in a better light is what I've named the "Universal Holistic Philosophy" or UHP and in this chapter we look at what are the detailed steps for learning to see the world from this broader and more open minded perspective. This is all while maintaining logic and a scientific minded skeptical inquiry on what is the truth.

Universal Holistic Philosophy

Universal Holistic Philosophy

2.1 Interests in the Unknown

The first requirement of a UHP thinker to have curiosity about the unknown and mysterious things which happen to you in your life or what you have learned which seems strange.

The moment in my life when this first occurred was when I read the book "Stranger Than Science" by Frank Edwards in the early 1960s. This book was full of strange and unexplainable stories and it really peaked my interest in all things unexplained. I think this led me on my life's quest of understanding the strange and weird things in this world of ours.

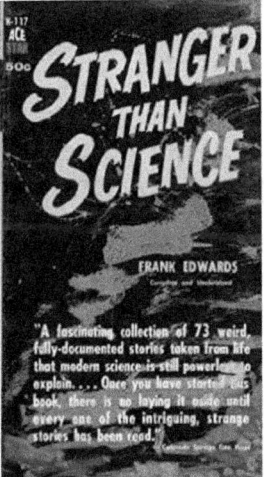

Not just accepting what you are taught in school is key. Validating what teachers and books tell you with your own inquiries is extremely important.

Universal Holistic Philosophy

Universal Holistic Philosophy

2.2 Science Fiction and Fantasy Are Important

Another vehicle I've found extremely useful is to read a large amount of Science Fiction and Fantasy books. I started with Comic Books and then graduated to full books as I grew into being a teenager.

Since that time I've read literally thousands of Sci Fi and Fantasy books which has exposed me to a huge number of concepts in the realm of all that I write about which includes some of these fields:

- Longevity & Immortality
- Super Intelligence
- Futuristic Technology
- Paranormal Abilities
- Aliens & UFOs
- Alternative Dimensions
- Time Travel
- Monsters and Cryptozoology
- Fantastic Realms
- Spiritual Evolution
- Unusual Social Developments in Society

I would also add any unusual concept you can think about is probably part of my reading history.

This long reading history has caused me to be open minded when I hear new concepts and to not throw them out because of previous biases, but to consider the logic of what I'm being told to see if the idea makes any kind of sense. Many very strange ideas have turned out to be true

Universal Holistic Philosophy

over the ages. You should view new ideas and concepts skeptically, but make sure your bias is not controlling your thinking.

Universal Holistic Philosophy

2.3 Getting a Good Education

Having a good education from Primary and Secondary schools, to University is also important. I admit I'm biased to my engineering education because this gave me a grounding in what conventional science and technology understands. This also includes the limits of what our Science and Engineering can't go beyond.

Next is part of a chapter from my first book titled "On Using the Scientific Method to Study the Paranormal" which reviews the potential limits of our scientific understanding:

Our Limited Understanding of Reality

Our scientific understanding of the world is only about 500 years old. Mankind has existed for over 100,000 years, and the universe is billions of years old.

Is humanity so arrogant as to say that we have a close to final understanding of the natural scientific laws of the universe, or should we be more humble and admit that we only understand a tiny fraction of what is out there, and much more is undiscovered than discovered.

I spent many years reading articles and journals from organizations like the ASPR (American Society for Psychical Research) who have done good experimental work for 60 years on validating and understanding psychic phenomenon.

However if I were to go to the average person on the street they would say that these things have never been proven.

Universal Holistic Philosophy

(I also read the standard scientific journals like Science and Scientific American.) Most scientists would also say that paranormal events haven't been proven to exist.

Instead of exploring how to understand and benefit from these abilities, most researchers in these areas are still being asked to prove that these things really exist. In this case many of the skeptics aren't really interested in the objective evidence because it would disrupt their cozy worlds.

The Scope of Reality

From the foregoing discussion on the scientific method and what is measurable, you can tell that I must have a significantly different idea of reality than the norm. Diagram #1 best illustrates my belief of our ability to understand Reality:

Universal Holistic Philosophy

The inner yellow circle represents what we can measure with our instruments today and perform experiments on to prove or disprove theories.

The red circle is a larger area, which we will eventually be able to measure to understand and prove or disprove the way things are

3) The blue outer circle is the largest area, and is that part of the universe which we may be able to experience but will never be able to measure and validate with objective scientific approaches.

We may be able to subjectively perceive a lot of things in the blue area, but will never have the tools and techniques to objectively quantify it.

This blue realm may also include such things as where the soul goes after death, the fundamental nature of God, and certain dimensions of space and time, which we can postulate but never prove or disprove.

The red region may be more amenable to creative approaches for objective measurement and validation.

However, there will have to be agreement among the scientific community on some new approaches, which may constitute legitimate standards for objective measurement of phenomena.

This may include indirect evidence, which is used in areas like particle physics.

Neutrinos for example can't be directly perceived, but their existence can be inferred by collisions with other particles, which make cloud tracks which we can directly perceive.

Universal Holistic Philosophy

The same approach should be transferable to validation of something like telepathy, where the medium of thought transference may not be understood at this point, but it can be validated through well-controlled blind studies and statistics.

I think that this type of validation issue of the objectivity of an experiment also presents barrier to further scientific progress.

Until new objectivity standards are set, we will never make good progress on a scientific understanding of consciousness and "non-physical" phenomenon.

It is very important to understanding how a new idea or concept will be evaluated by normal members of our society and to see what type of logic applies to any explanations of strange or unusual things which you might run across.

My technical education has also prepared me to better judge when a description of something new is a genuine effect or just a fraudulent or ignorant description of something which the person is trying to describe. Informed skepticism is a good quality to nurture.

Universal Holistic Philosophy

2.4 The Importance of an Open Mind

Having an Open Mind may seem obvious to many of you readers because you want to learn something new. However, you need to understand that in our world most people are propagandized from birth about what they should believe. You may actually be caught inside of one of these belief paradigms which will limit your ability to be open to new concepts.

Here are a few questions I request you consider to get outside of the artificial boundaries you may have already setup inside your heads:

- Is Man Made Global Warming real or just a political paradigm of people who want to increase their own money and powers?
- Do Aliens and UFOs really exist?
- What can you really to do with your life which others tell you is impossible?
- What limits have you imposed on yourself?
- Is the United States really as free as we think it is?

Universal Holistic Philosophy

- Does a spirit of God exist?
- What do insane people think?
- Does a duplicate of you exist somewhere else in the Universe?
- Is our destiny fixed or can we change it?
- Can people develop God Like Powers?
- Could you survive in the wilderness alone?
- Do You Question What you are Taught?
- Is the Bible all spiritual truth or should you also think outside of that box?

Universal Holistic Philosophy

2.5 Internet Resources

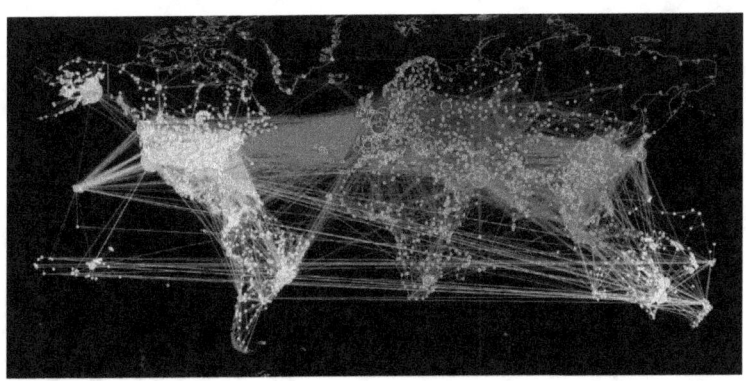

When I was growing up in the nineteen sixties and seventies, the only way to learn new facts was to go to a library or to school. And unless you lived near a good sized library there was a lot of knowledge around the world you would never learn or find out about.

Our access to the Internet in the last twenty years has really changed that paradigm so that almost anything you want to know is available online. Sure, you have to use your judgement since much of what is there could be false, but the plethora of information on any subject is many orders of magnitude greater than used to be possible just twenty years ago. (I use around the year 2000 as the year the information paradigm shifted since lots of information was just getting onto the internet before that.)

This makes researching almost any subject much easier and quicker than was possible to previous generations. I would almost say that Universal Holistic Philosophy was not possible before the invention of the Internet.

Universal Holistic Philosophy

Universal Holistic Philosophy

2.6 Expansion of the Scientific Method

The scientific method as used so far in history is still being used in a limited manner. To explore new paradigms which are still mysteries we need to use new techniques of the scientific method which have not been previously considered worthwhile. Here is more on some proposed alterations to using more subjective measurements on hard to study systems again from my book "On Using the Scientific Method to Study the Paranormal":

Subjective Versus Objective Knowledge

The scientific method is by its nature objective. In other words someone can independently verify all scientific observations with the right equipment and procedures.

This objective knowledge provides a strong foundation for building other science knowledge and technology, which makes our daily lives easier.

Subjective knowledge on the other hand is knowledge, which only the observer has. It has not been verified independently.

A good example of subjective knowledge is someone who has a strong vision of Jesus. The observer may feel that it was a real experience, but they usually can't prove it to someone through objective experimental results because only the observer was aware of the vision.

Can everything be measured Objectively?

Today we have so much confidence in the results of technology from application of the scientific method that

Universal Holistic Philosophy

hard scientific advocates assume that everything can be measured by experiment and proven true or false.

The logical extension of this reasoning for most people is that if something can't be measured, it doesn't exist.

Now of course many people believe in God, and they know God can't be measured. Even many scientists profess belief in God.

This belief is not logical if God can't be measured. Right?

However, scientists also admit that there are many clearly physical phenomena which they can't measure either.

Examples might include:

1) What is going on at the center of the earth?
2) Are the fundamental constants of the Universe the same at a point several light years from us?
3) What happened to the Universe in the first instant of the Big Bang?

These are questions which may never be answered with objective scientific observation

This limitation is basically one of instrumentation. If you don't have an instrument to measure something you can't experiment on it to generate objective results.

An example would be that there was no way to measure radiation when the understanding of radiation was too limited to have already developed instruments to measure it.

Universal Holistic Philosophy

This is an old problem which is part of what science is all about. Instruments have to be developed to make observations with the proper accuracy to prove or disprove a hypothesis.

There is a further problem with measurement:

Are there phenomena which exist in the world of consciousness which we don't haven't instruments to measure?

The anecdotal evidence is that the answer is yes—many people think they have experienced certain phenomena, (like telepathy) but we don't have any standard objective instruments to measure if it exists or not.

Here is another question to think about:

Are there events of consciousness which we can never develop traditional instruments for because they exist outside of our physical world?

The answer to this line of reasoning now becomes clearer—that there certainly are phenomena which we don't presently have the instruments today to properly measure. (How do you measure Love or Hate in a person?)

Even more discomforting than this is that there may be phenomena which we will never be able to measure even though we have a pretty good subjective idea that they may exist.

(How do you measure a ghost or a vision of the future—or the first instant of the Big Bang which created the Universe?)

Universal Holistic Philosophy

We have become very confident late in the 20th Century that we are close to understanding reality and close to developing an integrated theory of all the physical forces.

We think we are close to final answers in the sciences and soon being able to shut the door on ignorance and phenomena, which we can't verify objectively.

It's ironic that some scientists at the end of the 19th century also thought that they had discovered everything and there was nothing new to learn—just like the general feeling at the end of the 20th century.

Maybe the problem is that our vision of reality as proven by the scientific method is too narrow, and until we expand the scope of our thought to include a way of understanding other phenomena not so easily measured; we will never really get to a breakthrough understanding of reality.

We should also all have less hubris and more humility in our estimation of how far science has come and how far it has to go.

We can use a group of people to validate certain measurements as long as there are consistent controls about what they are observing. The group consensus uses their perceptions to answer questions being asked. Human perception of reality goes way beyond our scientific instruments.

2.7 Data Consistency

I also consider the consistency about different reports to be very important in determining the validity of data. A good example is Out of Place Artifacts.

There are many out of place artifacts I've found in researching my books. Two books these cases are from are "Strange Objects Which Should Not Exist" and "More Out of Place Artifacts". One group of findings in particular has caught my attention:

- Gold Thread-320 to 360 Million Years Old
- Ancient Screw-300 to 320 Million Years Old
- Iron Pot from Oklahoma-312 Million Years Old
- Mysterious Bell-298 to 320 Million Years Old
- A Screw Object Hidden Inside Stone 300 Million Years Old
- Ancient Modern Tools-300 Million Years Old
- Mechanical Part-300 Million Years Old
- Wheel in Coal Mine- 300 Million Years Old
- France-Tools in Rock 300 Million Years Old

Universal Holistic Philosophy

The above items really interest me because the examples cover a variety of objects found in coal or other strata and they all come from the same period in time of 300 to 320 million years old.

These age of these objects indicate evidence for the idea that an intelligent civilization existed on Earth at that time. There may be many other civilizations in Earth's history which we now have no records or artifacts about.

Universal Holistic Philosophy

2.8 Personal Experiences

Subjective personal experiences are also very important. Remember you are trying to determine the truth of mysterious subjects first to yourself then to others.

So things like paranormal or spiritual experiences which are very subjective will still be by helpful by adding to removing illusions and making apparent the truths you seek.

Many of my paranormal experiences reviewed in Chapter three of this book are good examples as to how subjective personal experiences can help individual understandings of time and other aspects of reality.

Universal Holistic Philosophy

Universal Holistic Philosophy

2.9 Allowing Time for Ideas to Develop

As I will describe in the next chapter the insightful conclusions I've come to on topics like Longevity and Ancient History didn't happen all at once. These realizations took time and happened gradually. If you go into researching a new topic and expect to determine an enlightened truth immediately then you are bound to be disappointed.

You have to build your narrative on a new and innovative topic over time as you continue to research the topic, develop new insights, and get feedback from others.

I took me about five years after I first started studying long lived people until I had developed a solid paradigm on the reasons for longevity and what one can do to improve human longevity.

Universal Holistic Philosophy

Universal Holistic Philosophy

2.10 Diverse Facts and Their Integration

The most innovative thinking I think I've accomplished was in my work on Longevity.

First, I started with the evidence I collected on long lived persons around the world.

Then I collected information over time from diverse articles and books on lifestyles, diets, psychological issues, and more.

My own experiences with premonitions caused me to realize that for a healthy person they would still be subject to accidents and premonitions are a way to mitigate that problem.

This all led to my development of "The 10 Principles of Personal Longevity" which is an integrated or Holistic Philosophy of living to improve the length of an individual's life by decades.

Universal Holistic Philosophy

Universal Holistic Philosophy

2.11 Thinking like Einstein

Einstein had an open minded, childlike, and confident approach to science-and as a result is considered one the greatest minds of all time. Here are some of the things he did:

Have no fear: If one had to point to a single quality that really set Einstein apart from the pack, it would be personal courage and self-confidence. Not only was he fighting against the old Newtonian guard with its concepts of absolute space and time, but he was fighting against making intellectual leaps without enough ballast.

Keep it simple: One of Einstein's famous quotes says, "Everything should be made as simple as possible, but no simpler", and he certainly followed his own advice, especially during his early career. James Clerk Maxwell had found a constant speed for light – 186,000 miles per second. But 186,000 miles per second relative to what? The next few years were spent constructing ingenious experiments trying to find the all-pervasive ether, until the Michelson-Morley experiment showed that there was no such medium. *Einstein simply took this result at face value*, and it became one of only two basic postulates of special relativity; the other one being that the laws of physics are the same in all inertial frames.

Do thought experiments: There was no other scientist who mastered the art of achieving insights through thought experiments as well as Einstein. From his wondering as a 16-year-old how the world would look like if he rode on a beam of light to his thinking about the equivalence principle (the equivalence of inertial mass with gravitational mass) after imagining free fall in an elevator, Einstein never ceased to imagine.

Focus on the ends, not the means: For a physicist, mathematics is a tool to achieve her goals. It can be a nerve-rackingly beautiful tool, but it's a tool nonetheless.

Young Einstein was the supreme embodiment of this attitude. He wasn't technically accomplished as a mathematician compared to fellow theoretical physicists like Dirac, Heisenberg, Chandrasekhar or Bethe. But where he surpassed every other physicist was in his physical intuition; he saw the physics first, and the mathematics came later.

Laugh through it all: Einstein had a bitingly sarcastic, self-effacing sense of humor. When he was thinking about the implications of his mass-energy equivalence, he wrote to a friend and wondered whether the "good Lord might be leading him around the nose."

Einstein's approach has a lot of similarities to what I'm talking about-enjoy your free thinking and find creative ways to measure, experiment, and justify your conclusions.

Understanding Reality is fun.

Universal Holistic Philosophy

3.0 Holistically Derived Understandings

Many of the ideas and concepts I've written about in my books are based on my own research and synthesis of ideas from my experiences and the diverse reading and research I've done. While I would like to think much of what I'm proposing is true, it may take decades or centuries to find general agreement with the ideas I propose. Of course some of my beliefs have been developed by others too. Mentioning all of these areas shows how my approach gets to these answers as part of my own sequence of efforts.

3.1 Investigating the Paranormal

One of my oldest interests which goes back to my teenage years is my interest in the paranormal. Was it real? What was it all about? How could I learn to have those experiences? I learned a lot about spiritual and psychic development in college and here is the process of how I learned to meditate and some of the results from my book titled "My Incredible Paranormal, Spiritual, and Out of the Box Experiences":

In the fall of 1973 I was accepted to Rensselear Polytechnic Institute which was my number one school-- a top nationwide technical school in upstate New York. I had been admitted to work on a degree for Physics. (My acceptance at Princeton was as a backup school but I decided not to go there.) At RPI I was thrust into an environment with lots of other kids who had also been at the top of their High School classes. I took courses in Calculus, Physics, Engineering, and much more. Learning in depth Mathematics and different levels of Physics including relativistic physics was really neat and I was very happy that I was finally getting a chance to learn what science knew about the world.

I also attended various free lectures on the paranormal and much more. At one of these lectures I met a very interesting man who would later become a Mentor to me. This Mentor's name was Sam Lentine. (Now deceased) Sam was middle aged, married with two kids and was working on his P.H.D. in Physics. He was also blind and a strong clairvoyant powerful healer. We became fast friends. Sam was planning on holding a meditation and psychic development class and I definitely want to attend. I helped Sam with some of the class details and attended the class which met weekly. Sam's meditation class was a turning point in my life. Whereas previously I thought I might have experienced paranormal phenomena, I was shocked to experience real abilities and phenomena as a result of Sam's class. One of the first things the students learned was how to do mantra meditation.

Universal Holistic Philosophy

In addition we were learning how to open the Crown Chakra on the top of the head. This is one of the seven main Chakras on the human body as described by Indian philosophers and masters.

The experience of meditation itself was all about concentrated relaxation. That when you focused on a mantra or were being led into the imagery of relaxation then your mind slowed down and calmed down in an unusual way. It was like your body was asleep but you were hyper aware of yourself and everything around you.

When the meditation was over you also felt well rested. Over a period of months and years of regular meditation it changes your personality to be living in the moment much more. The deeply relaxed meditative state became something I looked forward to each day as a way to tune the body and center myself in the spirit.

As I learned to meditate I also kept focusing on my crown chakra opening and energy coming into the top of my head. I was continually visualizing this happening as part of the techniques I was learning. After a couple of weeks I started feeling heat energy entering my head and moving down my body like liquid fire. This was amazing—and it was real. It wasn't like trying to convince myself that I had real clairvoyant abilities to read cards psychically. I could really feel the energy. As it moved through me it gradually helped open additional chakras. It was also somewhat destabilizing because I was really feeling the effects of eastern meditation techniques which western sciences said were impossible. Some classmates also experiencing the same energy effects became unstable, started taking drugs, and eventually dropped out of school. Was learning all of this meditation and additional Siddhis or abilities making me happier? Maybe it was. Between everything I was learning in school and learning meditation, vital forces, and how to heal it was certainly an exciting and generally fun time.

This learning about vital forces and meditation to get in touch with my spiritual consciousness became a passion for me. I was experiencing the best of both worlds. The world of science and the world of the spirit. I considered myself one of the luckiest guys alive. The reality of these vital forces was now undeniable to me. Some things you have to experience to learn that are real. The results of the meditations were amazing. I learned how to quiet my mind and became more sensitive to everything around myself.

Meditation also increased my inner peace which acted as a buffer against the stresses of the world. As for the energy I absorbed through my Crown Chakra, I could feel it energizing my body and it often flowed out through my hands when I wanted to heal anyone.

In fact this energy coursing through my body led me to do my first energy healings. While still in college I was at a Theosophy meeting in Schenectady, New York one evening when the leaders asked those with physical problems to come forward. Then they asked for healers. I came up to the front, put my hand on the shoulder of one guy who said he had painful shoulder problems and the energy flowed into him through my hands. Afterwards he said his shoulder didn't hurt anymore.

Years later I took a couple of Reiki healing courses and used to go to weekly Reiki sessions in Santa Monica, California for a couple of years where we did meditations and practiced healing each other.

Universal Holistic Philosophy

An example of what I learned was one thing Sam taught the class which was psychometry-how to sense people's energies on common objects we have like car keys and then tune into that person to tell things about them.

Even with these distractions, I still continued my studies and if anything, these psychic events made me even more fascinated with learning about the world around me.

One time in my room I decided to practice an invisibility experiment I had read about from the Rosicrucians during High School. Now that I knew how to meditate and manipulate vital forces I should be able to do it.

First I laid on my bed and went into a meditative state. Then I directed vital forces up to the ceiling of my room and used my arms and hands to direct the flow. I kept visualizing that the energy was creating a blue cloud. After doing this for twenty minutes or so I actually saw the blue cloud appearing. It was real. It was not imaginary. I stopped the exercise because I became afraid. The rest of the exercise was supposed to be to draw the blue cloud to surround my body. Then when I went among people I would be invisible to them. I figured I would do this another day, but the experiment validated to me that these vital forces energies existed and could be enabled by focus and will power.

One day I was at a table with other students and a girl I knew. I offered to hold her keys and tell things about her. She wasn't too excited about the prospect but I insisted. I ended up telling her several personal items about herself but instead of her being excited I could tell she was getting quite scared. I realized I needed to be more careful about these abilities since they might worry or cause concerns in most people.

Universal Holistic Philosophy

Universal Holistic Philosophy

3.2 The Nature of Time

The true nature of time has always been a subject of rigorous philosophical and scientific debate. I have looked at time from numerous different perspectives. These perspectives include:

- My Premonitional and Prophecy experiences
- University Courses in Relativistic Physics and Quantum Mechanics
- Research on individual stories of time travel for my book "Real Time Travel Stories".
- My Meditational and Spiritual events with our core Spirit which exists in a timeless and space less state.
- Time Travel science fiction stories

These points of view have given me more data to have a better understanding of time overall. There is still a lot I don't know about this topic but here are some things I've learned:

3.2.1 The Future has many probable Paths

I've written a couple of books about my premonitions of the future. We have the freedom of will to choose our future but there is also "Momentum" towards futures for us which are most likely. Here are a couple of my visions which illustrate this point:

Universal Holistic Philosophy

Visions

During the summer of 1975 I had a summer CO-OP job at General Electric's Gas Turbine engineering group in Schenectady, NY

At this time I used to meditate at my desk during the lunch hour.

One day in early August I was meditating and thinking about a trip I was planning to Cape Cod. My mind was wandering as I was thinking about what I would do there. My thoughts went to what I would do at the beach.

All of a sudden, I had a blinding flash of a scene where I was in the surf at the beach, and a surfboard was coming towards me. Then a shock occurred and I was thrown out of my meditation and was wide-awake.

I thought that this was pretty weird, and mentioned this to a friend or two.

Universal Holistic Philosophy

Two weeks later I was walking on the beach on Cape Cod. I saw a couple of guys with surfboards and asked where I could rent one to give it a try.
They said they had an extra one and I could try it with them. (I had totally forgotten my meditation vision at this point)

I tried to get up on that board all day, and had some modest success, but I was also getting exhausted in the process.

I decided to try it again and fell off when a big wave hit me. Next thing I knew I was coming up to the surface and I saw the exact same scene from my meditation.

The board hit me hard in the chin and almost knocked me out. I staggered to the shore and the two guys I was with helped me to the hospital where they put 10 stitches and 2 sutures into my chin.

The question arises—would I have been able to avoid the accident if I had remembered my vision and not gone surfing?

Later experiences have convinced me that the future is a set of probabilities, and we have free will to decide our actions.

I also had an experience on that trip of being able to partially heal my wounds very quickly through a deep meditation and application of psychic healing techniques. However, I do still have a small scar on my chin from this accident.

Universal Holistic Philosophy

We can see our likely futures and we can change them too. I learned this from other experiences described later in this book

Warnings of Danger Detroit

In 1980 I was moving from Dekalb, Illinois to Rochester, NY between assignments at General Electric, Inc.

While staying with my cousin outside Detroit, I made arrangements one evening to meet an old RPI friend Steve. We decided to go into downtown Detroit to the newly completed Renaissance Center to eat dinner and look around.

The Renaissance Center was built near the water and surrounded by slums.

After dinner we were walking out through the lower level in an area that was all boarded up with nobody else there.

Suddenly, I had this strong urge to turn around and go to find a restroom. I stopped walking forward because the urge was so strong.

I tried to walk forward again and again a very strong urge came to turn around and go back into the main center where other people were. I remarked to Steve that I couldn't go forward—that something wouldn't let me.

Just then two black guys in trench coats appeared about 30 feet away from behind one of the foundation pillars we were about to walk past. They started walking towards us with smiles pasted on their faces.

Universal Holistic Philosophy

My friend Steve took off running back into the main area and after a moment or two I figured I didn't know what these guys were carrying under their coats, so I ran too.

In less than 30 seconds we were back in a populated area with Police present, and the two guys chasing us gave us smiles like "next time we'll get you" and took off going the other way.

I had previously always tried to pray to God for protection, and tried to give a subconscious message to my senses to warn me of danger.

I'm convinced that whatever sense or "angel" warned me that evening, I would have been killed or severely wounded if I had continued walking out of the complex with no warning.

Universal Holistic Philosophy

Universal Holistic Philosophy

3.2.2 Time Travel Time Warps

I didn't used to believe that any types of time travel were possible except for time slowing down for persons travelling at relativistic speeds as the Theory of Relativity allows.

Then I did a lot of research for my book "Real Time Travel Stories" and found many legitimate stories people experienced which indicates that some types of holes or gateways exist in the fabric of time that people can travel through. Also, that they can go both ways—to visit past times and to return.

Here are a couple of examples of time travel experiences from my book "Real Time Travel Stories" which show that in some places and times an actual time warp can exist:

Bold Street, Liverpool, England

Bold Street is a site with the most experiences of time slips so far discovered. The below stories indicate that there is some type of time warp on this street which some people pass through.

The Liverpool Time Slips and Mysterious Occurrences in Bold Street are numerous. This location seems to be some type of time portal. Several stories follow.

The subject of time has always intrigued us. Is it as set as we have always believed? Or does time loop back on itself, giving us a glimpse of a shadowy past out of the corner of our eye.

Was is just our imagination that made us believe we had seen an object or building change before our very eyes, and seem as though we had stepped back into the past? When this happens we usually shake our heads and put it down to imagination.

But over the last few decades, something strange has been happening in or near Bold Street, Liverpool England.

Universal Holistic Philosophy

Not just a glimpse of the past, but full immersion into the strange and mysterious world of English History, if only for a few moments at a time.

The strange thing about the Bold street time slips is the actual time and place they are set. In the following cases, the people involved do not go back really far, but seem to visit a particular decade or decades.

So far, most of the sightings have centered around the 1950s and '60s. This is strange in itself. Most time travel experiences seem to take the recipient back to the 18th or 19th century. But not in this case.

Are these people simply copying each other in their experiences, or are they genuinely taking a step back in time?

The answer to this has to take into account whether they are doing it deliberately to get noticed. In other words are the perpetuating a hoax?

Another explanation could be mass hallucination.
And last but not least, they really are experiencing this strange phenomena!

The most important point is, the very first person that had this experience, obviously totally believed in what he saw, heard and felt.

So, does time flow like a river? Or does it twist and turn, going forward then sweeping back, picking up historic events and placing them down in front of you, if only for a few moments?

Universal Holistic Philosophy

In this first tale, we find Frank and his wife out for a stroll in Liverpool town center. It is 1996.

His wife decided that she wanted to go and buy a book at Waterstone's the large book store, and they started to walk towards the area of the shop.

As they approached Bold Street, Frank decided to go to another shop first, but bumped into his friend, and stopped to chat in the street. His wife went ahead without him.
A few moments later, Frank said goodbye, visited his shop and turned to go back to meet his wife. After reaching Bold Street, he headed on towards the bookstore. As he approached, he glanced up and was surprised to see the name, Cripps above the door. As he was about to cross over to see what was going on, a van swept past him with the name Cardin's on the side. The van driver honked his old fashioned horn and drove past.

Universal Holistic Philosophy

Looking around, Frank suddenly realized that things were not quite what they should be. He looked at the cars driving past and realized that they were all old fashioned vehicles such as people would drive back in the 50's and 60's.

And then he noticed the people. Men were wearing hats and macs, and the women were dressed in head scarves, full skirts and had old fashioned hair styles such as women wore just after the war.

By this time, Frank was beginning to feel slightly freaked out. He carried on crossing the road and headed towards the store.

As he got closer he noticed in the window there were handbags, shoes, and umbrellas. Suddenly he saw a young woman looking up at the shop sign. She looked confused.

She was wearing modern clothes and as she saw him approaching, she smiled at him.

Universal Holistic Philosophy

Frank went into the shop, closely followed by the young woman. When they entered he was surprised and pleased to see that it had indeed turned back into a bookshop. The young woman smiled, shook her head and said, 'that was strange, I thought it was a new clothes shop!' then she walked away looking extremely puzzled.

This may sound an unlikely tale, but the odd thing about it is that Frank was, in fact, a former Police officer who was used to dealing in facts, and definitely wasn't the type of person who would believe in the paranormal.

Frank never stopped talking about it. Was this a time slip? Evidently Cripps was a women's shop that sold clothes and other goods decades before!

And Cardin's was also a well-known Liverpool firm that owned vans around the same time.

Universal Holistic Philosophy

The second story concerns a young girl by the name of Imogen. She had decided to go into Liverpool to buy her sister Abigail a few things for her new baby. Upon arriving she was happy to see a new MotherCare store that had opened up on the corner of Lord Street and Whitechapel.

She wandered around the store, and picked up a few baby items such as cardigans, baby bibs, and gloves. She was surprised to see how cheap the items were, but thought they were on offer as the store had just opened. Taking them to the counter, she tried to pay with her credit card. The staff member looked at her suspiciously, and went off to get the manager.

When she came back, she looked at the card and told Imogen that they didn't take cards. So, disappointed, Imogen went and put the items back as she hadn't any money with her.

When she got home, she told her mother what had happened. Her mother was surprised and really puzzled. 'That store closed years ago,' she said. 'There is a bank there now, in fact that's where I have my account'. Not believing her, Imogen took her mother back to the same place the next day. Sure enough the store wasn't there. It was a bank, just as her mother had told her.

The third tale is of a young man named Sean, who, while shoplifting in Liverpool back in 2006, ran away from a Security Guard and headed down Hanover Street. Trying to shake off the Guard, Sean, 19, turned into a dead end street called Brookes Alley.

By this time he was out of breath and started to get a tight sensation in his chest. He soon realized that actually it wasn't a problem with him, but the atmosphere around him.

He waited for the Guard to come around the corner after him, but he never appeared. So, thinking he had given him the slip, he sauntered back out and started to walk down Hanover Street again. But he soon realized that something was wrong.

The road looked different, and so did the pavement. He noticed cars driving by that looked very old fashioned, and the road works that he knew were there, were now gone. Soon he saw that the people around him were wearing strange clothes. Crossing over to Bold Street, he noticed that there were traffic lights where they weren't before, and bushes growing around the Lyceum, near a bar that he recognized.

He carried on walking. Soon he began to feel that something was not quite right. Then he began to panic. He realized that somehow he had stepped back in time. And the time slip was not going away.

Then he remember his Cell phone. Taking it out of his pocket, he tried to get a signal, but of course it didn't work. Eventually he began to really panic, but soon spotted a kiosk selling newspapers and headed over.

Leaning over the Stand, he took a look at the front page of the Daily Post. There in bold lettering was the date. 18th May 1967.

He wondered what to do. What happens if he can't get back to his own time? What about family and friends? So, speeding up his pace, he reached H. Samuel the Jewelers, and tried his phone once again. This time it

Universal Holistic Philosophy

worked. Sighing with relief he looked around and realized that he had returned to the present. But the strange thing was, he could still see, down the end of the road, people still walking around in 1967.

By this time Sean had seen enough, and dived onto a bus to go home. When he was interviewed by the local newspaper later, he stated over four times, the exact account.

Now, you may think that Sean was making the story up to escape from the guard. But the strange tale didn't end there. When the Security Guard was interviewed, he stated that when he ran after Sean, and turned down the dead end alley after him, he said that Sean had completely disappeared!

When the newspaper checked out the facts of Sean's story, they found that everything he said was historically accurate.

Universal Holistic Philosophy

These three stories are just the tip of the iceberg. There are many tales from around Liverpool that tell of time slips, ghosts and other strange phenomenon. The stories keep coming thick and fast, and of course the more tales, the more likely people will start to believe that they are all being made up, or as the saying goes, Urban Tales. So, what do you think? Real life time slips, imagination, mass hallucination or purely tales that have started out as fun but have turned into the greatest Urban Legends of all time.

My opinion is that, yes, something did happen.

Probably to the first guy, Frank who was just out shopping with his wife. The others? Maybe it was a case of mistaken roads, taking a wrong turning or just a glitch in the person's memory. By the time they get home they totally believe what happened.

Or is it true? There are so many cases concerning Bold Street, and just about anywhere else in Liverpool, that maybe, just maybe they are all living on top of the biggest time slip phenomena in the World.

Universal Holistic Philosophy

Sir Victor Goddard

Sir Victor Goddard's trip into the unexplained involved an airplane flight. This was a much more personally harrowing experience.

In 1935, while a Wing Commander, Goddard flew a Hawker Hart biplane to Edinburgh, Scotland, from his home base in Andover, England, for a weekend visit. On the Sunday before flying back, Goddard visited an abandoned airfield in Drem, near Edinburgh, this location being closer to his final destination than the airport at which he landed. The Drem airfield, constructed during the First World War, was a shambles. The tarmac and four hangars were in disrepair, barbed wire divided the field into numerous pastures, and cattle grazed everywhere. It was now a farm, and completely useless as an airfield.
On Monday, Goddard began the flight back to his home base. The weather was dark and ominous, with low clouds and heavy rain. Goddard was flying in an open cockpit over mountainous terrain without radio navigational aides

or cloud flying instruments. Rain began beating down on his forehead and onto his flying goggles badly which obscured his vision. He thought he could climb above the clouds, but he was wrong. He made it to 8,000 feet, looking for a break in the clouds. There was none.

Suddenly Goddard lost control of his plane. It began to spiral downward. He struggled with the controls. He could speed up or slow down, but he could not stop the spin. He was unsure of his location, but knew he was falling rapidly and might smash into the mountains before coming out of the clouds. The sky became darker, the clouds turning a strange yellowish-brown. The rain came down even more heavily. Goddard's altimeter showed he was only a thousand feet above the ground and dropping rapidly. At two hundred feet and still spiraling downward, he began to see a bit of daylight through the murky gloom, but his spiral toward seemingly inevitable death was far from over. Goddard was now flying at 150 miles per hour. He emerged from the clouds over "rotating water" that he recognized as the Firth of Forth. He was still falling.

Suddenly, he saw directly before him a stone sea wall with a path, a road, and railings on top of it. The road seemed to be slowly rotating from left to right. The cloud cover was down to forty feet. Goddard was now flying below twenty feet and was within an instant of tragedy. A young girl with a baby carriage ran through the pouring rain. She ducked her head just in time to avoid Hart's wingtip. Goddard succeeded in leveling out his plane after that. He barely missed striking the water after clearing the sea wall by a few feet.

He was now flying only several feet above a stony beach. Fog and rain obscured all distant visibility, but Goddard somehow located his position. He identified the road to

Universal Holistic Philosophy

Edinburgh and soon was able to discern, through the gloom, the black silhouettes of the Drem Airfield hangars ahead of him, the same airfield he had visited the day before. The rain became a deluge, the sky grew even darker, and Goddard's plane was shaken violently by the turbulent weather as it sped toward the Drem hangars-and into a different world.

Suddenly, the sky turned bright with golden sunlight. The rain and the farm had vanished. The hangars and the tarmac appeared to have somehow been rebuilt in a brand-new condition. There were four planes lined at the end of the tarmac. Three were standard Avro 504N trainer biplanes; the fourth was a monoplane of an unknown type-the RAF had no monoplanes in 1935. All four airplanes were bright yellow. No RAF airplanes were painted yellow in 1935. The airplane mechanics were wearing blue overalls. RAF mechanics never wore anything but brown overalls when working in hangars in 1935.

It took Goddard only an instant to fly over the airfield. He was only a few feet above the ground-just high enough to clear the hangars-but apparently none of the mechanics saw him or even heard his plane. As he sped away from the airfield, he was again engulfed by the storm. He forced his plane upward, flying at 17,000 feet and then, for a time, at 21,000 feet. He managed to return to his home base safely.

Goddard felt elated when he landed. He then made the mistake of telling fellow officers about his eerie experience. They looked at him as if he were drunk or crazy. Goddard decided to keep silent about what had happened to him. He did not want a discharge from the RAF on mental grounds.

Universal Holistic Philosophy

In 1939, Goddard watched as RAF trainers began to be painted yellow and the mechanics switched to blue coveralls. The RAF introduced a new training monoplane exactly like the one he had seen in his flight over Drem. It was called the Magister. He learned that the airfield at Drem had been refurbished.

Another twenty-seven years went by, but Goddard never forgot what had happened. He played it through over and over in his mind. It was not until 1966 that he wrote of this experience. Over the years he had become convinced that there was no way he could have known that the RAF would change the colors of their trainers and their mechanics' overalls four years before these changes took place. Goddard finally concluded that he must have glimpsed the future-or even traveled into it-for a brief moment in time.

Universal Holistic Philosophy

3.2.3 Our Core Spirit Sees Time Differently

The following Prophecy really taught me about the probabilities of the future, how to change them, and the momentum of the time track we are currently on:

Planning a Trip to Spain

During early August of 1998, my wife and I decided to send her and our kids to visit her mother in Barcelona, Spain.

I was going to buy a ticket separately, and meet them there during early September.

When I started to call the travel agent to book my ticket I had a terrible feeling of fear about taking the flight.

Universal Holistic Philosophy

I tried two other times to book the ticket during the week for a September 2nd departure, and each time I got the same strong feelings of fear and death.

I have always prayed and tried to guard myself mentally to avoid disasters, so finally I took the warning seriously and decided not to go at all.

This was very difficult to do since I really wanted to see my wife and kids, and this meant I would be home alone for a month.

Work wasn't an excuse either, since I wasn't doing any really heavy contract work at the time and could easily have taken the time off.

I called my wife and told her my decision, and she was surprised, but agreed for me to follow my instincts.

On September 2nd the Swissair disaster occurred on a plane leaving Kennedy airport in New York, which crashed in Newfoundland Canada with all lives lost.

I would not have originally been booked on that flight, but could have easily ended up on it since I was due to fly through Kennedy airport, and any delay might have caused me to switch planes. I will never know for sure, but this was a very strong warning.

I should also mention that for several years before this event I had strong feelings that my I would be killed in the near future. After this happened those feelings ended.

This story is a great illustration that the probabilities of the future can be changed but there is a momentum to events which makes it difficult to do so.

Universal Holistic Philosophy

My Conclusions about Time:

My thinking is that our core consciousness is intimately connected with time since the core spirit of our being lives in a timeless and space less realm. This explains how we can see probabilities of the future through our spirit.

Also, time warps do exist at certain places and times and allow people to travel through them and return. These warps may also somehow be enabled by our consciousness too.

Current Physics theories do not consider the evidence of these experiences so there is something missing from these theories which needs to be added. Scientists want to develop a Universal Theory of Everything but they will never do so as long as they ignore the spirit and paranormal events.

Universal Holistic Philosophy

3.3 Spiritual Enlightenment

Worldwide religions are very important to expose individuals to Spirituality but one has to dig much deeper to realize our own spiritual truths.

Lots of other persons studying spiritual enlightenment have come some to the same conclusions I have. My conclusions are from personal experiences so I don't expect to convince others of their validity.

I studied world religions and enlightenment for many years starting as a teenager. My learning meditational practices and developing paranormal abilities were side effects of the general spiritual development process. Then I had a couple of enlightenment experiences—which were very powerful. It is described in my book "The Enlightenment Experience".

Universal Holistic Philosophy

Here is My Heart Opening Experience:

In the summer of 2012 I had an amazing full blown heart opening experience. I had previously experienced a few occasional heart openings over the last couple of hearts from the "Heart Rose" exercise.

My most recent experiences put those previous ones to shame. In September of 2012 a close female friend of mine who I love dearly told me that she had always loved me unconditionally but not romantically.

Our discussion finally led me to conclude that intimacy was not in our future. However, this woman was someone I had a very deep spiritual connection to and who already had an open heart.

I think spending a lot of time with her over a couple of years plus some other training and experiences of mine had raised my vibrational level to a point where I was ready for a full heart opening.

Universal Holistic Philosophy

The process started when I reviewed my feelings for her in my own mind—where my ego was very hurt. Then I recalled all of the good times we had experienced and the spiritual healing and different she had made in my life. As I reviewed those experiences and realized how grateful I was for them happening was when my heart started to open… in a much greater way than ever before.

There was a lot of heat in my heart and I could feel the heat spreading throughout my body. I also felt more "connected" to everyone and everything than I was before. This experience went on for hours and happened to some extent everyday thereafter.

I might just be sitting on my couch watching TV and suddenly my heart would open and I would feel wonderful and like my body was being charged with a powerful energy…

-
Sometimes the energy collecting in my chest gets so powerful that I feel a lot of pressure like a heart attack. I may have to exercise or cool myself off to relieve the pressure. (It's definitely not a medical issue since the pressure quickly goes away and there are no other heart attack symptoms.)

I had learned at the age of nineteen to open my crown chakra and take in energy for healing but this heart experience was an order of magnitude more powerful—and just felt like a different and more personal form of energy. This experience has become the second major spiritual milestone in my life.

It felt-and still feels today like a drug sometimes where I'm getting high just sitting there and experiencing this natural spiritual life force in my life.

Universal Holistic Philosophy

As the last few months have progressed I am beginning to truly understand the spiritual experience that the tongues of fire are described as in the Bible:

ACTS 2:3 And there appeared to them tongues as of fire distributing themselves, and they rested on each one of them.

The significance of this quote is since I was experiencing that fire in my heart—a burning sensation which is pleasurable and doesn't really burn me.

Imagine just sitting there feeling tendrils of fire winding through the body in the bloodstream providing a very pleasant feeling and happiness. Words do not do the experience justice.

It's kind of like being on a drug—but with only good side effects

I've also realized that the opening is continuing—it's not a one-time maximum experience but an evolution of the spirit within my body.

Some other effects I've noticed is that people have started treating me differently—especially women. It's like they subconsciously sense the change in my energy and want to get close to me.

Although I'm a fairly attractive middle aged guy I'm not used to beautiful women I don't know starting conversations with me and initiating contact.

I'm also not used to a stranger at a restaurant counter reaching into her purse to change a twenty dollar bill of mine. I didn't ask her to do this—she just looked at me and did it when I asked for change at the counter.

Several other experiences come to mind—but you get the idea. Everyone seems to resonate with a person who has an open heart. They sense the spirit of God has come down into that person and want to be closer to it.
This energy of the spirit has also increased my healing ability and has improved my immune system so that I'm not experiencing the colds and flus of my family and the people around me. In previous years I would have caught all of these diseases.

This whole experience has been a great blessing in my life.

****

Now, for the first time in my whole life I truly realize from my experiences with heart opening that my ultimate happiness is inside of me, not in a relationship with somebody.

A relationship can add to the positive aspects of my life, but will not create a happy relationship in itself.

****

I also want to tell you about emotional effects from the heart opening since I find myself more connected to everyone and everything around me and the "hole" I used to have in my heart is gone.

This "hole" is something I've experience throughout most of my life—even after many years of spiritual development. I always felt that something was empty inside of me. I'm sure that this "lack" inside of me also affected my relationships—both friends and lovers—for many years.

Universal Holistic Philosophy

Now that the emotional "hole" has been filled I feel wonderful and want everyone to be able to enjoy what I've been so blessed to experience.

In talking with other spiritual persons who have experienced heart openings (most of them seem to be women) I've learned that quite a few of them have had the experience. Admittedly many of these women are already deep into spiritual practices. Maybe it's also because women are naturally centered closer to their heart than men. I don't know—but its interesting food for thought.
I do feel that I now have an extra dimension of experience in all types of relationships. The spiritual dimension of love and the connections it creates seems to overlay the emotional and intellectual relationship connections that used to be all I knew.
This allows me to be more centered in how I communicate with others—and not get bound up into my emotions but into my spiritual heart.

This spiritual dimension lets me experience other people more fully, and better see the oneness and perfection in all of us.

"One thing: you have to walk, and create the way by your walking; you will not find a ready-made path. It is not so cheap, to reach to the ultimate realization of truth. You will have to create the path by walking yourself; the path is not ready-made, lying there and waiting for you. It is just like the sky: the birds fly, but they don't leave any footprints. You cannot follow them; there are no footprints left behind." — Osho

All of my experiences led me to write the book titled "Pure Spirituality & God" which focuses on the concept that standard religions are important parts of our society and

Universal Holistic Philosophy

very useful to introduce people to spirituality, but one has to dig much deeper to find individual spiritual truths.

Universal Holistic Philosophy

Universal Holistic Philosophy

3.4 Immortality and Longevity

My adventures in exploring Immortality and Longevity have been really fascinating and occurred over a period of at least five years.

Back in 2008 I was interested in writing about a subject which would be of real interest to me and other readers. The subject I choose was Longevity because there were always stories in the press about persons who claimed to have lived into their mid one hundreds or longer and people who the history books said had also lived very long lives. Yet the prevailing attitude in society was that nobody could live much beyond the age of 100 years and even that was very rare.

So I did a lot of research through books I purchased and on the Internet. I found to my shock that there are many recorded cased of persons living well into their mid one hundreds and some living over 200 years.

(I also found that birth certificates didn't even come into general usage in the United States until World War Two when many people were applying for positions which needed a security check.)

Next was to try to offer an explanation of how this might be possible since the evidence of long lived persons was overwhelming. I reviewed many writings on spirituality and longevity which provided some anecdotal evidence to lots of things people can do for themselves to live even decades longer. This led to my first book on the subject titled "Physical Immortality: A History and How to Guide" in 2009.

Universal Holistic Philosophy

Following this I continued to research Longevity and Immortality which led me to determine that I could write a set of principles of what people could learn to improve their longevity in a systematic way. Here are the 10 Principles of Personal Longevity I developed. This is detailed in the book by the title "The 10 Principles of Personal Longevity":

The 10 Principles of Personal Longevity

1) Real Long Lived Persons Exist
People really have lived a long time-so you can do it too
2) Define Your Purpose in Life
Know your life purpose-To live life with meaning
3) Enable Your Life Urge
Know without doubt that you will live a long and happy life
4) The Importance of a Spiritual Connection
A spiritual connection is important for happiness & long term health
5) Having Love in your Heart
Unconditional Love is is real-It will make you happier and healthier
6) Activate your Vital Forces
Improve the vitality of your energy body for health and to enjoy life more
7) The Science of Longevity
Use new therapies and discoveries from Science & Medicine
8) Keep your Physical Body Healthy
Eat a proper diet, use herbal supplements, and exercise
9) Use Your Intuition for Safety
Learn to use your intuition to keep you safe
10) Implement the above principles in your life
Implement these principles for long term health, greater happiness, and extended longevity

---

After this I developed an online training program to teach coaches about the 10 Principles and how to teach their clients how to live them.

Universal Holistic Philosophy

I have kept writing books on Longevity and have learned a lot more over the last 10 years which I've also updated into my online training program.

A side effect of marketing the Physical Immortality book was an online discussion forum about the book which led a person nick named "Stealth" to approach me and declare he is actually 2,800 years old. I will not go into all the details here but what I learned from these conversations made a great book titled "The Commentaries of Living Immortals". Whether true or not it is a good story and my open mindedness may have led to new insights about real living immortals.

I do strongly believe that the 10 Principles approach to longevity I developed was based on my unbiased analytical look at the lives on long lived persons and what we do know about what enhanced longevity. It was by combining all this information together that created this approach to helping people live longer and healthier lives.

Universal Holistic Philosophy

3.5 Ancient History

Several years ago I watched a video online about a new potential location for the mythical city of Atlantis to be in the Richat Structure in West Africa. This video made a real impression on me so I did a lot of research and came to the conclusion that this is the correct site of the fabled Atlantis. The resulting book has a lot of information which makes a good case for this being the real site of Atlantis. The book is titled "The Real Atlantis: In the Eye of the Sahara".

My interest was piqued about ancient civilizations and this became the first of eleven books on this topic. I found many old ruins which push back the dates of civilization much further than archeologists and anthropologists would ever agree are valid. My next two titles were "Ancient Civilizations", and "Ancient Civilizations-Book Two".

One of the themes of ancient human history is the existence of Giants in the world. Some are described in the

Universal Holistic Philosophy

Bible like Goliath, and King Og among others. What I read about giants led me to believe they have existed throughout history all over the world. My next book was on Giants and titled "The History of Antediluvian Giants" which covered the history of Giants in recorded history and from archeological digs.

While doing my research I also noticed a lot of articles on "Out of Place Artifacts". In other words, finds of intelligently built objects like bells, hammers, and mortars/pestles found in rock strata millions of years old. Impossible right?

Anyway, this is also a fascinating subject so I researched and wrote two more books titled "Strange Objects Which Should Not Exist" and "More Out of Place Artifacts".

I was gradually building more background knowledge of ancient civilizations and relics. As I did so I started to have some epiphanies. My next realization was that there must have been a great disaster which destroyed civilization about 10,500 B.C. This realization occurred from putting together various scientific, geological, archeological, and other writings from several sources. Here is more information from "The Destruction of Civilization About 10,500 B.C."

This information all fits together to provide my theory on how, when, and why, ancient civilizations fell. This information is described next:

Universal Holistic Philosophy

3.5.1 Younger Dryas Comet Impacts

The Younger Dryas impact hypothesis (YDIH) or Clovis comet hypothesis posits that fragments of a large (more than 4 kilometers in diameter), disintegrating asteroid or comet struck North America, South America, Europe, and western Asia around 12,800 years ago. Multiple airbursts/impacts produced the Younger Dryas (YD) boundary layer (YDB), depositing peak concentrations of platinum, high-temperature spherules, meltglass, and nanodiamonds, forming an isochronous datum at more than 50 sites across about 50 million km2 of Earth's surface. Some scientists have proposed that this event triggered extensive biomass burning, a brief impact winter and the Younger Dryas abrupt climate change, contributed to extinctions of late Pleistocene megafauna, and resulted in the end of the Clovis culture.

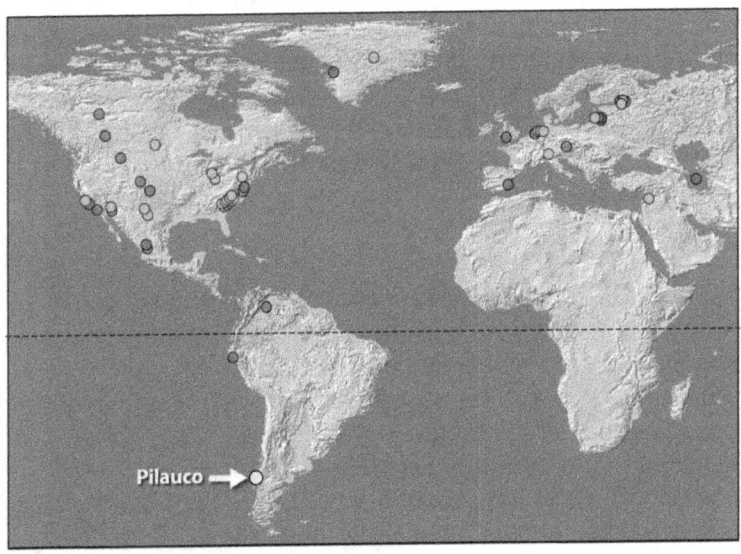

Above is a map from Mario Pino et al. 2019 showing 53 Younger Dryas boundary sites. Orange dots represent 28

sites with peaks in both platinum (Pt) and other impact proxies such as high-temperature Fe-rich spherules. Red dots represent 24 sites with impact proxies but lacking Pt measurements.

21st Century Comet Revival and Elaboration

In 2006, this hypothesis was revived in The Cycle of Cosmic Catastrophes: How a Stone-Age Comet Changed the Course of World Culture, a trade book by Richard Firestone, Allen West and Simon Warwick-Smith published by Inner Traditions – Bear & Company and marketed in the category of Earth Changes. It proposed that a large air burst or earth impact of one or more comets initiated the Younger Dryas cold period about 12,900 BP calibrated (10,900 14C uncalibrated) years ago.

In May 2007, at a meeting of the American Geophysical Union in Acapulco, Firestone, West, and around twenty other scientists made their first formal presentation of hypothesis. Later that year, the group published a paper in the Proceedings of the National Academy of Sciences (PNAS) that suggested the impact event may have led to an immediate decline in human populations in North America at that time.

In 2008, C. Vance Haynes Jr. published data to support the synchronous nature of the black mats, emphasizing that independent analysis of other Clovis sites was required to support the hypothesis. He was skeptical of the bolide impact as the cause of the Younger Dryas and associated megafauna extinction but concluded "... something major happened at 10,900 YBP (14C uncalibrated) that we have yet to understand." The first debate between proponents and skeptics was held at the 2008 Pecos Conference in Flagstaff, Arizona.

Universal Holistic Philosophy

In 2009, a paper in the journal Science asserted that nanodiamonds were evidence for a swarm of carbonaceous chondrites or comet fragments from air burst(s) or impact(s) that set parts of North America on fire, caused the extinction of most of the megafauna in North America, and led to the demise of the Clovis culture. A special debate-style session was convened at the 2009 AGU Fall Meeting in which skeptics and supporters alternated in giving presentations.

In 2010, astronomer William Napier presented evidence that fragments of a comet—initially 50 to 100 kilometers in diameter—could have been responsible for such an impact, and that the Taurid complex is formed of the remaining debris. Napier refined this model and published further research in 2019.

In 2012, another paper in PNAS offered evidence of impact glass that resulted from the impact of a meteorite.

Another group of scientists reported evidence supporting a modified version of the hypothesis—involving a fragmented comet or asteroid—was found in lake bed cores dating to 12,900 YBP from Lake Cuitzeo in Guanajuato, Mexico. It included nanodiamonds (including the hexagonal form called lonsdaleite), carbon spherules, and magnetic spherules. Multiple hypotheses were examined to account for these observations, though none were believed to be terrestrial. Lonsdaleite occurs naturally in asteroids and cosmic dust and as a result of extraterrestrial impacts on Earth. Lonsdaleite has also been made artificially in laboratories.

In 2013, scientists reported a hundredfold spike in the concentration of platinum in Greenland ice cores that are dated to 12,890 YBP with 5 year accuracy. They attribute this platinum anomaly to the likely impact of a large (~0.8 km) iron-rich meteorite locally onto Greenland's ice, which would have been "unlikely to result in an airburst or trigger wide wildfires proposed by the YDB impact hypothesis."

But they write that the large Pt anomaly "hints for an extraterrestrial source of Pt". An alternative suggestion is that the Greenland Pt anomaly was caused by a small local iron meteorite fall without any widespread consequences, but this is disputed by the paper's authors who claim that a global platinum anomaly is expected due to the ~ 20 year lifetime of the platinum signal.

In 2016, a report on further analysis of Younger Dryas boundary sediments at nine sites found no evidence of an extraterrestrial impact at the Younger Dryas boundary.

Also that year, an analysis of nanodiamond evidence failed to uncover lonsdaleite or a spike in nanodiamond concentration at the YDB. Radiocarbon dating, microscopy of paleobotanical samples, and analytical pyrolysis of fluvial sediments "[found] no evidence in Arlington Canyon for an extraterrestrial impact or catastrophic impact-induced fire." Exposed fluvial sequences in Arlington Canyon on Santa Rosa Island "features centrally in the controversial hypothesis of an extra-terrestrial impact at the onset of the Younger Dryas."

In 2017, scientists reported a Pt anomaly dating at eleven continental sites dated to the Younger Dryas, which is linked with the Greenland Platinum anomaly.

In 2018, some researchers interpreted the undated Hiawatha Glacier impact crater in Greenland as evidence for the Younger Dryas impact event due to its location.

Universal Holistic Philosophy

Two papers were published dealing with an "extraordinary biomass-burning episode" associated with the Younger Dryas Impact.

In 2019, scientists reported evidence in sediment layers with charcoal and pollen assemblages both indicating major disturbances at Pilauco Bajo, Chile in sediments dated to 12,800 BP. This included rare metallic spherules, melt glass and nanodiamonds thought to have been produced during airbursts or impacts. Pilauco Bajo is the southernmost site where evidence of the Younger Dryas impacts has been reported. This has been interpreted as evidence that a strewn field from the Younger Dryas impact event may have affected at least 30% of Earth's radius. Also in 2019, analysis of age-dated sediments from a long-lived pond in South Carolina showed not just an overabundance of platinum but a platinum/palladium ratio inconsistent with a terrestrial origin, as well as an overabundance of soot and a decrease in fungal spores associated with the dung of large herbivores, suggesting large-scale regional wildfires and at least a local decrease in ice age megafauna.

Universal Holistic Philosophy

3.5.2 The Vulture Stone

Gobekli Tepe was first discovered in the 1990s and is now recognized as one of the most important historical sites in the world. It is the same location as that for the first farming of grains and domestication of animals. It is a potential location for the start of civilization.

The ancient site of Gobekli Tepe in Turkey has recently been a source of excitement about a column in one of the temple ruins which seems to indicate a large disaster occurred in 10,950 B.C. when comets struck the Earth. Here is a picture of the column:

Gobekli Tepe

Ancient stone carvings confirm that a comet struck the Earth around 11,000 B.C., a devastating event which wiped out woolly mammoths and sparked the rise of civilizations.

Universal Holistic Philosophy

Experts at the University of Edinburgh analyzed mysterious symbols carved onto stone pillars at Gobekli Tepe in southern Turkey, to find out if they could be linked to constellations.

The markings suggest that a swarm of comet fragments hit Earth at the exact same time that a mini-ice age struck, changing the entire course of human history.

Scientists have speculated for decades that a comet could be behind the sudden fall in temperature during a period known as the Younger Dryas. But recently the theory appeared to have been debunked by new dating of meteor craters in North America where the comet is thought to have struck.

However, when engineers studied animal carvings made on a pillar – known as the vulture stone – at Gobekli Tepe they discovered that the creatures were actually astronomical symbols which represented constellations and the comet.

The idea had been originally put forward by author Graham Hancock in his book "Magicians of the Gods".

Using a computer program to show where the constellations would have appeared above Turkey thousands of years ago, they were able to pinpoint the comet strike to 10,950BC, the exact time the Younger Dryas begins according to ice core data from Greenland.

The Younger Dryas is viewed as a crucial period for humanity, as it roughly coincides with the emergence of agriculture and the first Neolithic civilizations.

Universal Holistic Philosophy

Before the strike, vast areas of wild wheat and barley had allowed nomadic hunters in the Middle East to establish permanent base camps. But the difficult climate conditions following the impact forced communities to come together and work out new ways of maintaining the crops, through watering and selective breeding. Thus farming began, allowing the rise of the first towns.

Edinburgh researchers said the carvings appear to have remained important to the people of Gobekli Tepe for millennia, suggesting that the event and cold climate that followed likely had a very serious impact.

11,000 B.C. is the correct timeframe which Plato says that Atlantis was destroyed. Plato's words were:

> *But afterwards there occurred violent earthquakes and floods; and in a single day and night of misfortune all your warlike men in a body sank into the earth, and the island of **Atlantis** in like manner disappeared in the depths of the sea.*

A cometary disaster hitting around North America could also cause huge tidal waves all over the Earth. If the comet hit part of the Atlantic Ocean, the tidal wave could have been a mile high or more after moving across the Ocean and to Africa. Such a wave could have moved inland and drowned the entire city of Atlantis.

Given the uncertainties in radio carbon dating and other scientific methods having dates of these occurrences within a thousand years of each other could indicate events which happened at the same time.

From Critias this quote:

Universal Holistic Philosophy

When afterwards sunk by an earthquake, became an impassable barrier of mud to voyagers sailing from hence to any part of the ocean.

This indicates that the channel to the Ocean from the former city of Atlantis was clogged with mud after the catastrophe. We don't know what type of ruins might lie beneath the soil of the Eye of the Sahara, because no excavations have been done.

Universal Holistic Philosophy

3.5.3 The Flood of the Bible

There are many myths worldwide of a flood which destroyed human civilization.

The flood-myth motif occurs in many cultures as seen in: the Mesopotamian flood stories, manvantara-sandhya in Hinduism, the Gun-Yu in Chinese mythology, Deucalion and Pyrrha in Greek mythology, the Genesis flood narrative, Bergelmir in Norse mythology, flood during the time of Nuh (Noah) of Qur'an, the arrival of the first inhabitants of Ireland with Cessair in Irish mythology, in parts of Polynesia such as Hawaii, the lore of the K'iche' and Maya peoples in Mesoamerica, the Lac Courte Oreilles Ojibwa tribe of Native Americans in North America, the Muisca and Cañari Confederation in South America, Africa, and some Aboriginal tribes in Australia.

Earliest Comet Based Versions of the Great Flood

Universal Holistic Philosophy

Painting from 1840 depicting a comet causing the Great Flood. The Eve of the Deluge, by John Martin, 1840. Depicts a comet causing the Great Flood.

The original hypotheses about a comet impact that had a widespread effect on human populations can be attributed to Edmond Halley, who in 1694 suggested that a worldwide flood had been the result of a near-miss by a comet. The issue was taken up in more detail by William Whiston, a protégé of and popularizer of the theories of Isaac Newton, who argued in his book A New Theory of the Earth (1696) that a comet impact was the probable cause of the Biblical Flood of Noah. Whiston also attributed the origins of the atmosphere and other significant changes in the Earth to the effects of comets.

This hypothesis was subsequently popularized by Minnesota congressman and writer Ignatius L. Donnelly in his book Ragnarok: The Age of Fire and Gravel (1883), which followed his better-known book Atlantis: The Antediluvian World (1882). In Ragnarok, Donnelly argued that an enormous comet struck the Earth approximately 12,000 years ago, destroying an advanced civilization on the "lost continent" of Atlantis. Donnelly, following Halley and Whiston, attributed the Biblical Flood to this event, which he hypothesized had also resulted in catastrophic fires and significant climate change. Shortly after the publication of Ragnarok, one commenter noted, "Whiston ascertained that the deluge of Noah came from a comet's tail; but Donnelly has outdone Whiston, for he has shown that our planet has suffered not only from a cometary flood, but from cometary fire, and a cometary rain of stones."

Universal Holistic Philosophy

3.5.4 More Ancient History Realizations

Another category of my books is on Aliens and UFOs, and I was looking for an explanation of how intelligently made objects could be found so far back in time when humanity didn't even exist yet. There have been too many items found which rules out the ideas that they were all hoaxes.

This led to my next title "A Theory of Ancient Prehistory and Giant Aliens". In this book I provided a lot of evidence of recorded giant footprints and other things like wheel ruts which were in fossilized rocks. Again, I was just continuing to build on my knowledge of Ancient History as an outside observer of the subject. My knowledge of Ancient Aliens also supported the idea that these out of place relics were really from ancient intelligent alien civilizations which existed well before humanity came into the picture.

The last book came about when I realized that there is no timeline going all the way from the eras of these out of place artifacts and ancient ruins which are not dated in many cases all the way to modern times. This book is a major synthesis of all of my ancient civilizations works and some Alien UFO historical information to produce the book

Universal Holistic Philosophy

"A Timeline of Intelligent Life on Earth". Although some other Authors have also proposed ancient aliens on Earth I think I'm the first one to tie it all together into the outline of a timeline of intelligent life which has existed on the Earth. Here is a graphic I designed to help illustrate this idea:

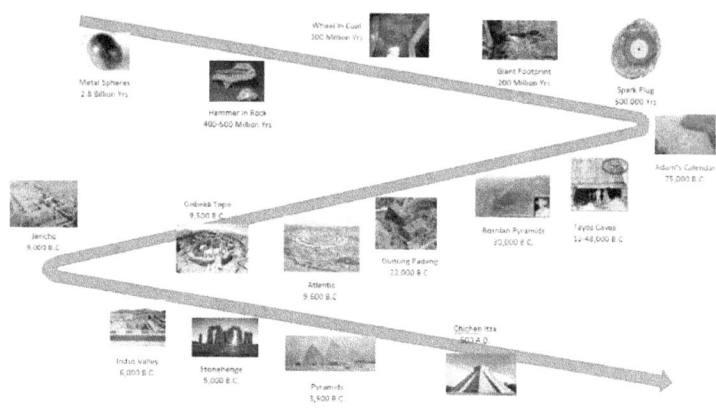

Universal Holistic Philosophy

3.6 UFOs and Aliens

UFOs and Aliens is a subject of great interest to many people and so I thought I would investigate this subject too since my paradigm is to investigate mysteries and try to determine the truth.

There are many UFO and Alien books on the market so I'm not really providing any new information overall. But my books emphasize the stories and logic in many of the other ones about the ideas that Aliens and UFOs have been here on Earth all through our history and also thousands if not millions of years before mankind existed. In fact one title of mine is "Four Evidences for Aliens and UFOs in Earth's History" which looks at the existence of these beings on Earth from multiple perspectives to offer more validity of their really having been here.

One of my books on this subject is also titled "Aliens Are Already Among Us" which provides another set of validations on this topic.

Universal Holistic Philosophy

I also want to mention that several years ago I attended the large "Contact in the Desert" Alien/UFO event in the California Desert near Joshua Tree National Park. Thousands of people attended and there were many lectures by notables in the UFO and Alien community.

Unfortunately, I have to say that nine out of ten of the presenters seemed like nut cases because they didn't really present any evidence to build their viewpoints or were obviously mentally disturbed. But there were a few speakers who had a lot of logic in their presentations. I guess that should be expected in a subject this controversial.

Universal Holistic Philosophy

3.7 The Multiverse

A lot of physicists have theorized that we live in a Multiverse. In other words that there are additional dimensions which differ from ours. These dimensions could have different physical properties like ours such as time running at different rates.

Some think that other dimensions are created when a decision is made here and the branches of these timelines create new dimensions. Maybe yes, or maybe no. But other dimensions would answer a lot of questions and provide answers to mysteries.

My interest was to see if there was any truth to the occasional stories I've read about people coming from or going to other dimensions. I went into this research pretty skeptical that there was any truth to the idea of alternative dimensions but kept an open mind. I was looking for consistency in stories to check the validity of the claims.

Again to my great surprise I found a lot of stories which if true point specifically to the validity of other dimensions existing and people being able to pass back and forth between them. Here are a couple of those stories from my book "Stories of Parallel Dimensions":

Universal Holistic Philosophy

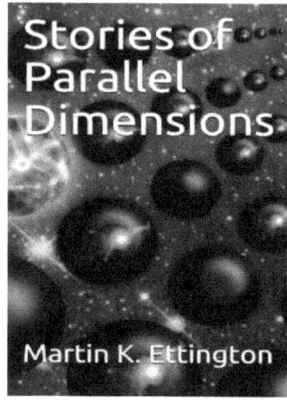

Here are some of the best interdimensional stories I found for this book:

A vanishing road

(Here is a story told in the first person) About 4 years ago, I lived in this fairly small flyspeck of a town. At the time, I had lived there for about 12 years so I knew my way around. Our house was about a mile and a half away from the nearest neighborhood.

Universal Holistic Philosophy

Our mom intentionally picked that house due to the lack of neighbors. It was tucked away on a back road, with the woods surrounding it. Every now and again, I liked to take walks with my little brother, who at the time was about 13.

We decided to do just that. We headed up the road and decided to try to find a new path or a new clearing that we hadn't discovered yet, when we noticed something a little shocking. Just off the road that led almost directly to the neighborhood, there was a brand new paved road. Every road in that part of town was a gravel road, so seeing an out of place paved road was pretty unusual. We stared at it for a while, and came to the conclusion that it must have been made within the last few days, due to the modern but slow growth of the town. However, we had no explanation for how they did it so fast.

We decided to explore it a bit. I remember as soon as we set foot on the road, the air became notably colder, by at least 5 degrees. The road itself was a black pavement, but no dividing lines. It was surrounded by some thick, red trees that resembled redwoods, but they were too short and non-native to our state (southern Arkansas). We walked on the road for about 3 miles until we decided to head back due to it getting dark. When we got off the road, we felt the temperature go back up. My brother and I agreed to explore it the next day.

At roughly noon the following day, we set back out to explore this place, only to discover that the entire road was now missing. When I say missing, I mean the trees that were cleared to make it had apparently grown back, with no sign of the redwood-like trees. We even began to explore the woods once more, but only to find no sign that it ever existed. When we asked our parents about it, they

Universal Holistic Philosophy

said they knew nothing about any new road work being done near us.

The Money

This one comes from Reddit user "Shadowjack00". The story is actually from a friend of theirs. They said -A friend of mine from Toronto was in bar in New York on a business trip. As he paid for drinks and food a few times during the night, he noticed a blond woman watching him intently. After a while, she came over and struck up a conversation with him. Eventually she commented that she noticed he had colored money in his wallet.. she seemed, as he put it, eager to the point of being afraid to mention it.

He told her yes, he was Canadian and had some Canadian money in his wallet. She demanded he show her and he did. She was apparently very disappointed and started to cry a bit but refused to talk about it further.

Feeling bad, he tried to cheer her up and the two of them got a bit hammered together. (He admitted that his goal was to not spend the night alone but he did anyway). At one point he asked her why she wanted to see his Canadian money and, after a lot of coaxing, and more drinks, she told him. She said that up until a few months before, US money had been different colors. Ones were green, tens were blue and hundreds were brown. He couldn't remember the other colors she told him. She said the colors were not bright like Canadian money but were sort of a set of dark tints. She said the brown hundreds were called "bricks" because the brown tint they had was similar to a brick. Then, she said, one day they were all green and she was the only one who seemed to remember them being different colors. My friend pressed her and she said there were other differences too, not just money that she noticed. Popular TV programs were different or had the wrong actors in the lead roles. There was more but my friend couldn't remember what else she said. After a while, she just started crying saying she finally thought she had found someone who would convince her she wasn't going crazy and he turned out to be a Canadian instead. Shortly after, she left and that was the end of it.

Universal Holistic Philosophy

3.8 Human Will and Survival

I learn something new from every book I write. The one I wrote titled "33 Incredible Survival Stories" really taught me about the power of the human will to survive.

I had thought I knew a lot about survival from my training as an Eagle Scout, but hadn't realized how important an individual's will to survive was as part of the story.

From the book on survival stories here is one of the best stories on the expedition of Sir Earnest Shakelton:

Universal Holistic Philosophy

Ernest Shakelton's Voyage to the South Pole

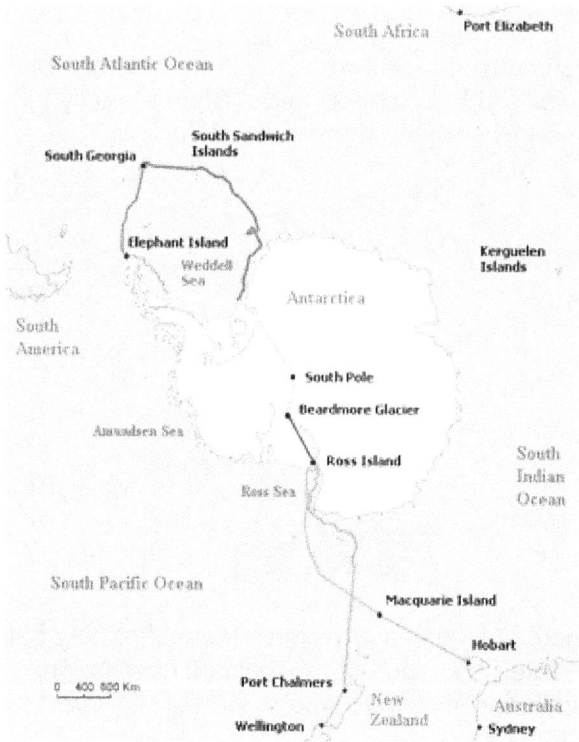

Imperial Trans-Antarctic Expedition, 1914–1917

Preparations

Main articles: Imperial Trans-Antarctic Expedition, Ross Sea Party, and List of personnel of the Imperial Trans-Antarctic Expedition

Outline of Antarctica coast, with different lines indicating the various journeys made by ships and land parties during the expedition

Universal Holistic Philosophy

Shackleton published details of his new expedition, grandly titled the "Imperial Trans-Antarctic Expedition", early in 1914. There is a legend that says Shackleton's newspaper article was written a certain way so that he could better narrow down and select candidates for his expedition. Two ships would be employed; Endurance would carry the main party into the Weddell Sea, aiming for Vahsel Bay from where a team of six, led by Shackleton, would begin the crossing of the continent. Meanwhile, a second ship, the Aurora, would take a supporting party under Captain Aeneas Mackintosh to McMurdo Sound on the opposite side of the continent. This party would then lay supply depots across the Great Ice Barrier as far as the Beardmore Glacier; these depots would hold the food and fuel that would enable Shackleton's party to complete their journey of 1,800 miles (2,900 km) across the continent.

Shackleton used his considerable fund-raising skills, and the expedition was financed largely by private donations, although the British government gave £10,000 (about £900,000 in 2019 terms). Scottish jute magnate Sir James Caird gave £24,000, Midlands industrialist Frank Dudley Docker gave £10,000, and tobacco heiress Janet Stancomb-Wills gave an undisclosed but reportedly "generous" sum. Public interest in the expedition was considerable; Shackleton received more than 5,000 applications to join it.

His interviewing and selection methods sometimes seemed eccentric; believing that character and temperament were as important as technical ability, he asked unconventional questions. Thus physicist Reginald James was asked if he could sing; others were accepted on sight because Shackleton liked the look of them, or after the briefest of interrogations. Shackleton also loosened some traditional hierarchies to promote

Universal Holistic Philosophy

camaraderie, such as distributing the ship's chores equally among officers, scientists, and seamen. He also socialized with his crew members every evening after dinner, leading sing-alongs, jokes, and games. He ultimately selected a crew of 56, twenty-eight on each ship.

Despite the outbreak of the First World War on 3 August 1914, Endurance was directed by the First Lord of the Admiralty, Winston Churchill, to "proceed", and left British waters on 8 August. Shackleton delayed his own departure until 27 September, meeting the ship in Buenos Aires.

Crew

While Shackleton led the expedition, Captain F. Worsley commanded the Endurance and Lieutenant J. Stenhouse the Aurora. On the Endurance, the second in command was the experienced explorer Frank Wild. The meteorologist was Captain L. Hussey, also an able banjo player. McIlroy was head of the scientific staff, which included Wordie.

Alexander Macklin was one of two surgeons and also in charge of keeping the 70 dogs healthy. Tom Crean was in more immediate charge as head dog-handler. Other crew included James, Hussey, Greenstreet, a carpenter Henry McNeish, and a biologist named Clark. Of later independent fame was the photographer Frank Hurley, known on this mission for his perilous shots.

There was a (male) cat named Mrs Chippy that belonged to the carpenter Henry McNeish. Mrs Chippy was shot when the Endurance sank, due to the belief that he would not have survived the ordeal that followed.

Universal Holistic Philosophy

Loss of Endurance

Endurance departed from South Georgia for the Weddell Sea on 5 December, heading for Vahsel Bay. As the ship moved southward navigating in ice, first year ice was encountered, which slowed progress. Deep in the Weddell Sea, conditions gradually grew worse until, on 19 January 1915, Endurance became frozen fast in an ice floe.

On 24 February, realising that she would be trapped until the following spring, Shackleton ordered the abandonment of ship's routine and her conversion to a winter station. She drifted slowly northward with the ice through the following months. When spring arrived in September, the breaking of the ice and its later movements put extreme pressures on the ship's hull.

Shackleton after the loss of Endurance

Until this point, Shackleton had hoped that the ship, when released from the ice, could work her way back towards Vahsel Bay. On 24 October, water began pouring in. After a few days, with the position at 69° 5' S, 51° 30' W, Shackleton gave the order to abandon ship, saying, "She's going down!"; and men, provisions and equipment were transferred to camps on the ice. On 21 November 1915, the wreck finally slipped beneath the surface.

For almost two months, Shackleton and his party camped on a large, flat floe, hoping that it would drift towards Paulet Island, approximately 250 miles (402 km) away, where it was known that stores were cached. After failed attempts to march across the ice to this island, Shackleton decided to set up another more permanent camp (Patience Camp) on another floe, and trust to the drift of the ice to take them towards a safe landing.

Universal Holistic Philosophy

By 17 March, their ice camp was within 60 miles (97 km) of Paulet Island; however, separated by impassable ice, they were unable to reach it. On 9 April, their ice floe broke into two, and Shackleton ordered the crew into the lifeboats and to head for the nearest land.

After five harrowing days at sea, the exhausted men landed their three lifeboats at Elephant Island, 346 miles (557 km) from where the Endurance sank. This was the first time they had stood on solid ground for 497 days. Shackleton's concern for his men was such that he gave his mittens to photographer Frank Hurley, who had lost his during the boat journey. Shackleton suffered frostbitten fingers as a result.

Open-boat journey

Launching the James Caird from the shore of Elephant Island, 24 April 1916 Elephant Island was an inhospitable place, far from any shipping routes; rescue by means of chance discovery was very unlikely. Consequently, Shackleton decided to risk an open-boat journey to the 720-nautical-mile-distant South Georgia whaling stations, where he knew help was available. The strongest of the tiny 20-foot (6.1 m) lifeboats, christened James Caird after the expedition's chief sponsor, was chosen for the trip. Ship's carpenter Harry McNish made various improvements, including raising the sides, strengthening the keel, building a makeshift deck of wood and canvas, and sealing the work with oil paint and seal blood.

Shackleton chose five companions for the journey: Frank Worsley, Endurance's captain, who would be responsible for navigation; Tom Crean, who had "begged to go"; two strong sailors in John Vincent and Timothy McCarthy, and finally the carpenter McNish. Shackleton had clashed with

Universal Holistic Philosophy

McNish during the time when the party was stranded on the ice, but, while he did not forgive the carpenter's earlier insubordination, Shackleton recognized his value for this particular job. Not only did Shackleton recognize their value for the job but also because he knew the potential risk they were to morale. This allowed for Shackleton to remain in control of the morale of his crew members. The attitudes of his men were a point of emphasis in leading his men back to safety.

Shackleton refused to pack supplies for more than four weeks, knowing that if they did not reach South Georgia within that time, the boat and its crew would be lost. The James Caird was launched on 24 April 1916; during the next fifteen days, it sailed through the waters of the southern ocean, at the mercy of the stormy seas, in constant peril of capsizing. On 8 May, thanks to Worsley's navigational skills, the cliffs of South Georgia came into sight, but hurricane-force winds prevented the possibility of landing. The party was forced to ride out the storm offshore, in constant danger of being dashed against the rocks. They later learned that the same hurricane had sunk a 500-ton steamer bound for South Georgia from Buenos Aires.

On the following day, they were able, finally, to land on the unoccupied southern shore. After a period of rest and recuperation, rather than risk putting to sea again to reach the whaling stations on the northern coast, Shackleton decided to attempt a land crossing of the island. Although it is likely that Norwegian whalers had previously crossed at other points on ski, no one had attempted this particular route before. For their journey, the survivors were only equipped with boots they had pushed screws into to act as climbing boots, a carpenter's adze, and 50 feet of rope. Leaving McNish, Vincent and McCarthy at the landing point on South Georgia, Shackleton travelled 32 miles (51 km) with Worsley and Crean over extremely dangerous mountainous terrain for 36 hours to reach the whaling station at Stromness on 20 May.

The next successful crossing of South Georgia was in October 1955, by the British explorer Duncan Carse, who travelled much of the same route as Shackleton's party. In tribute to their achievement, he wrote: "I do not know how they did it, except that they had to — three men of the heroic age of Antarctic exploration with 50 feet of rope between them – and a carpenter's adze".

Rescue

Shackleton immediately sent a boat to pick up the three men from the other side of South Georgia while he set to work to organize the rescue of the Elephant Island men. His first three attempts were foiled by sea ice, which blocked the approaches to the island. He appealed to the Chilean government, which offered the use of the Yelcho, a small seagoing tug from its navy. Yelcho, commanded by Captain Luis Pardo, and the British whaler Southern Sky reached Elephant Island on 30 August 1916, at which point the men had been isolated there for four and a half months, and Shackleton quickly evacuated all 22 men. The Yelcho took the crew first to Punta Arenas and after some days to Valparaiso in Chile where crowds warmly welcomed them back to civilization.

There remained the men of the Ross Sea Party, who were stranded at Cape Evans in McMurdo Sound, after Aurora had been blown from its anchorage and driven out to sea, unable to return. The ship, after a drift of many months, had returned to New Zealand. Shackleton travelled there to join Aurora, and sailed with her to the rescue of the

Universal Holistic Philosophy

Ross Sea party. This group, despite many hardships, had carried out its depot-laying mission to the full, but three lives had been lost, including that of its commander, Aeneas Mackintosh.

Persons who are determined and think ahead have a great chance to survive a disaster, and their positive attitude is what makes in possible in many cases.

Universal Holistic Philosophy

Universal Holistic Philosophy

3.9 Legendary Animals and the Multiverse

I was curious about claims there were many cryptozoological animals in the world so I decided to do an overall survey of these legends and try to determine which ones had some real truth to them. This led to my book "Are Cryptozoological Animals: Real or Imaginary?" This effort turned into several other books on specific crypto animals like Dragons, Thunderbirds, or Sea Monsters.

Separately I've researched many alternative dimension stories. This has led me to an interesting idea. The stories of many of these animals seem very real and the reports are consistent. But where would these an animals hide on our Earth?

So maybe we should also consider that many of these cryptids don't originate on our Earth but are from parallel dimensions. This would account for them often appearing

Universal Holistic Philosophy

from nowhere and then disappearing again to be unseen to a long time.

A good case can be made that Thunderbirds are cross dimensional creatures because there is no place known in which they could normally exist, but there are definite sightings of them around the United States and have been for hundreds of years.

Universal Holistic Philosophy

3.10 Our Malleable Universe

I've had many paranormal experiences in my life and some pretty cool spiritual enlightenment experiences too. This has given me a good understanding of the spiritual development process and the development of paranormal abilities as a side effect.

I'm also an Engineer with a lot of Physics mixed into my education so I have a pretty good understanding of the conventional views of engineering and science as taught in our universities.

My fascination with the unusual and unknown also led me to write about off the wall subjects like the paranormal, time travel, and moving between dimensions. The research I did convinced me that these strange and weird phenomena do exist and are not fantasies.

My books on time travel and dimensions include:

- Real Time Travel Stories
- The Real Nature of Time: An Analysis of Physics, Prophecy, and Time Travel Experiences
- Stories of Parallel Dimensions
- We Live in a Malleable Reality-and We Can Change It

Universal Holistic Philosophy

My numerous experiences and research have led me to conclude that although we may think we live in a stable Universe—that is a big misunderstanding. The more I learn the more convinced I am that the Universe is much more complicated than we ever thought.

In other words, my view of our Universe is constantly evolving as I learn more about it. A few years ago I would have considered my current viewpoints to be ludicrous and gullible.

My viewpoints on these subjects would not have changed if my research on time travel, parallel dimensions, the paranormal, spirituality, and more was not very objective and if I included common belief biases. My thinking about our reality is constantly evolving as I have more experiences and information.

4.0 Directions for Education

Our education in colleges and universities today is too narrow and thought restricting. I maintain that in engineering and other technical schools there should be courses on "Out of the Box Thinking" in additional to the traditional technical curriculum.

Students should be exposed to ideas and concepts they might consider impossible with supporting evidence for those concepts where feasible.

Many of the examples I cite in this book would be good in a course like that. The main point is to bring up subjects which are normally dismissed by technical people to show that with the proper point of view people might learn some real understandings to these mysteries.

There also needs to be more of a synthesis of information from different professions to find the truth of many curious phenomena.

Here is an example of how technologies from different professions modified the science of archeology and anthropology to a large degree.

Guessing the age of finds in digs by archeologists and anthropologists used to be really guessing game. The best dates came from looking at similar items which had known dates to compare against.

Universal Holistic Philosophy

In 1946 Carbon 14 dating started to measure the level of radioactive Carbon 14 in objects. It was first used to measure the age on known objects. Later as the usage expanded to more objects Carbon 14 dating became a popular tool for those digging up ancient history to use as a reliable dating tool.

Later, more types of radioactive dating techniques were used for different materials. These additional techniques span short and very long periods of time. These include the following types of measurements:

- Rubidium-Strontium dating (Rb-Sr) ...
- Potassium-Argon dating (K-Ar) ...
- Samarium-Neodymium (Sm-Nd) ...
- Rhenium-Osmium (Re-Os) system. ...
- Uranium-Lead (U-Pb) system. ...
- The SHRIMP technique. ...
- Fission track dating.

Universal Holistic Philosophy

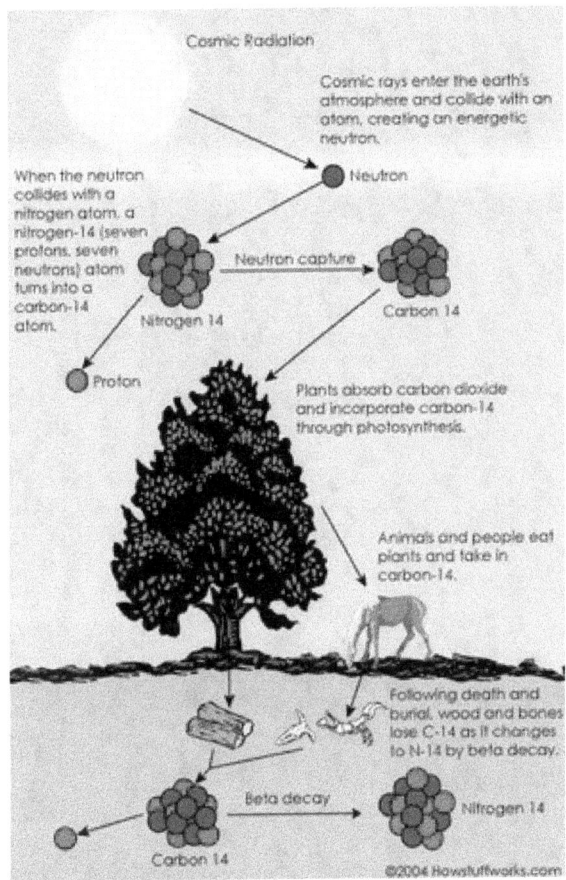

How Carbon 14 Aging Works

Overall, this is a great example of how specific types of new technology enhanced several existing professions.

The more people learn about diverse subjects the more they can combine knowledge in one area to leverage overall knowledge in other areas of study.

Universal Holistic Philosophy

Universal Holistic Philosophy

5.0 Summary

As our civilization becomes more sophisticated and we learn more about different subjects we still have many mysteries which are considered pseudoscience by many serious technical professionals.

The solution to many of these mysteries will come from taking the approaches recommended in this book using Universal Holistic Philosophy.

Scientists and Engineers need to stop rejecting ideas because they aren't comfortable with them. The greatest discoveries often cause the most dramatic improvements and disruptions in society.

It takes a lot of courage to hold an unpopular viewpoint, even if you are correct.

Wishing you all the best in your future creative endeavors.

Martin K. Ettington
October 2021

Universal Holistic Philosophy

6.0 Bibliography

1. https://3quarksdaily.com/3quarksdaily/2021/10/how-to-think-like-albert-einstein.html. *How To Think LIke Einstein.* [Online]

2. Ettington, Martin K. *On Using the Scientific Method to Study the Paranormal.* 2000.

3. —. *My Incredible Paranormal, Spiritual, and Out of the Box Experiences.* 2021.

4. —. *Use Intuition and Prophecy To Improve Your Life.* 2020.

5. —. *The Enlightenment Experience.* 2013.

6. —. *33 Incredible True Stories of Survival.* 2020.

7. —. *Real Time Travel Stories.* 2020.

8. —. *A Timeline of Intelligent Life on Earth.* 2021.

9. —. *Stories of Parallel Dimensions.* 2021.

10. —. *STrange Objects Which Should Not Exist.* 2020.

11. —. *More Out of Place Artifacts.* 2021.

12. —. *The Destruction of Civilization About 10,500 B.C.* 2021.

13. —. *We Live in a Malleable Reality-And We Can Change It.* 2021.

www.ingramcontent.com/pod-product-compliance
Lightning Source LLC
Chambersburg PA
CBHW071515220526
45472CB00003B/1033